NATURE'S FREAK SHOW: UGLY BEASTS

THE FRIGHTFUL PROBOSCIS MONKEY

BY JANEY LEVY

Gareth Stevens
PUBLISHING

Please visit our website, www.garethstevens.com. For a free color catalog of all our high-quality books, call toll free 1-800-542-2595 or fax 1-877-542-2596.

Library of Congress Cataloging-in-Publication Data

Names: Levy, Janey, author.
Title: The frightful proboscis monkey / Janey Levy.
Description: New York : Gareth Stevens Publishing, [2020] | Series: Nature's freak show : ugly beasts | Includes index.
Identifiers: LCCN 2019017331| ISBN 9781538246108 (paperback) | ISBN 9781538246122 (library bound) | ISBN 9781538246115 (6 pack)
Subjects: LCSH: Proboscis monkey--Juvenile literature.
Classification: LCC QL737.P93 L48 2020 | DDC 599.8/62--dc23
LC record available at https://lccn.loc.gov/2019017331

First Edition

Published in 2020 by
Gareth Stevens Publishing
111 East 14th Street, Suite 349
New York, NY 10003

Copyright © 2020 Gareth Stevens Publishing

Designer: Katelyn E. Reynolds
Editor: Monika Davies

Photo credits: Cover, p. 1 Berendje Photography/Shutterstock.com; cover, pp. 1-24 (curtain background) Africa Studio/Shutterstock.com; cover, pp. 1-24 (wood sign) Rawpixel.com/Shutterstock.com; cover, pp. 1-24 (marquee signs) iunewind/Shutterstock.com; p. 5 David Evison/Shutterstock.com; p. 7 Desmond Morris Collection/UIG via Getty Images; p. 9 Sergey Uryadnikov/Shutterstock.com; p. 11 Xiao Mai/China News Service/VCG via Getty Images; p. 13 Mohd KhairilX/Shutterstock.com; p. 15 dikobraziy/Shutterstock.com; p. 17 Jerry Redfern/LightRocket via Getty Images; p. 19 SIA KAMBOU/AFP/Getty Images; p. 21 Loop Images/UIG via Getty Images.

All rights reserved. No part of this book may be reproduced in any form without permission in writing from the publisher, except by a reviewer.

Printed in the United States of America

Some of the images in this book illustrate individuals who are models. The depictions do not imply actual situations or events.

CPSIA compliance information: Batch #CW20GS: For further information contact Gareth Stevens, New York, New York at 1-800-542-2595.

CONTENTS

Presenting the Proboscis Monkey 4
Why the Fleshy Proboscis? 6
Monkey Mates and Mothers 8
Raising Babies 10
Gathering in Groups..12
Home and Habitat..14
Tasty Leaves 16
Hunted 18
Endangered...20
Glossary...22
For More Information...23
Index 24

Words in the glossary appear in bold type the first time they are used in the text.

PRESENTING THE PROBOSCIS MONKEY

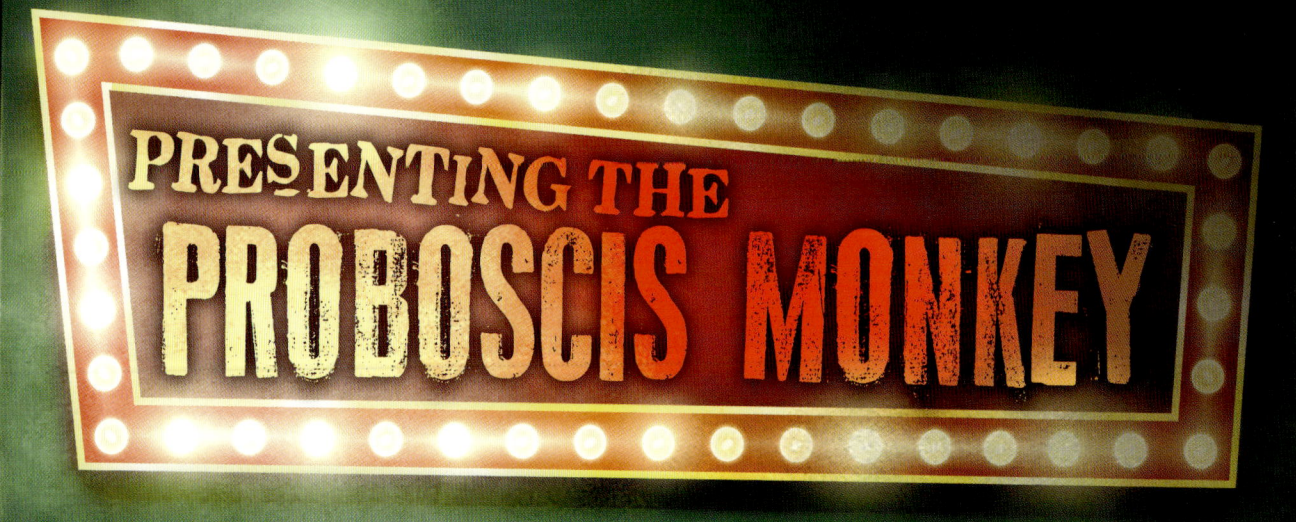

You might've seen lively monkeys playing around at the zoo. Or, perhaps you've visited places where monkeys live. Monkeys are **primates,** like people. Their faces even look a little like human faces. But the male proboscis monkey's face has an unusual feature.

"Proboscis" means "long, thin nose," and male proboscis monkeys have a fleshy nose that hangs over their mouth. People might find it strange or funny. But females find the males' nose quite lovely! You'll learn about these monkeys inside this book.

THE PROBOSCIS MONKEY IS SOMETIMES KNOWN AS THE LONG-NOSED MONKEY.

WHY THE FLESHY PROBOSCIS?

You don't have a long, fleshy nose that hangs down over your mouth. So, why does the male proboscis monkey have this kind of nose?

Male proboscis monkeys with bigger noses are bigger altogether. Female proboscis monkeys are drawn to male monkeys with bigger noses because they believe bigger monkeys will make better **mates**. Also, a male proboscis monkey with a larger nose can let out a louder mating call, which lets female monkeys know he's bigger—and likely a better mate.

STRANGE BUT TRUE! A nose hanging down over your mouth can make eating quite tricky! Male proboscis monkeys often have to push their nose to the side when they're eating.

THE NOSE OF THE MALE PROBOSCIS MONKEY CAN MEASURE OVER 4 INCHES (10 CM) LONG.

MONKEY MATES AND MOTHERS

Among proboscis monkeys, females are often the ones to choose their mate and decide when it's time to mate. They let males know it's mating time by drawing up their mouth in a kiss and shaking their head from side to side. Mating season runs from February until November each year.

After mating, the female has a single baby about 5 1/2 months later. The baby is usually born at night while the mother sits on a tree branch. That doesn't sound very comfortable!

STRANGE BUT TRUE! Male proboscis monkeys weigh up to about 50 pounds (23 kg), while females weigh up to 26 pounds (12 kg).

MALE

FEMALE

BABY

IT'S EASY TO TELL MALE AND FEMALE PROBOSCIS MONKEYS APART. NOT ONLY DO MALES HAVE A HUGE NOSE, BUT THEY'RE ALSO *MUCH* BIGGER THAN FEMALES.

RAISING BABIES

Adult proboscis monkeys have a pinkish face. Their light brown fur turns a reddish color around their head and shoulders. Their arms, legs, and tail are colored gray. Baby monkeys look quite different. They have dark fur and blue faces!

Like human babies, proboscis monkey babies are helpless when they're born. Their mothers carry them, nurse them, and keep them clean. Babies eat their first solid food when they're about 6 weeks old. They stay with their mother for over a year.

STRANGE BUT TRUE! Male monkeys don't care for their babies the same way female monkeys do. But males do help keep babies safe by chasing danger away.

BY THE TIME A BABY IS ABOUT 2 ½ MONTHS OLD, ITS FACE WILL CHANGE TO A GRAY COLOR. LATER, THE BABY'S FACE WILL BECOME THE SAME PINKISH COLOR AS ADULT FACES.

GATHERING IN GROUPS

Proboscis monkeys often gather in two kinds of groups. The first group is made of a single male plus several females and their young. The second is a **bachelor** group made of only male monkeys. Unlike other monkeys, proboscis monkeys don't always stay with the same group. They may switch to another group, mainly in their early years.

At night, several groups often come together to sleep in large gatherings called bands. They may also form bands to travel together during the day.

STRANGE BUT TRUE! A mother proboscis monkey has lots of help. Other female monkeys in the group will help carry and care for her baby.

A GROUP OF PROBOSCIS MONKEYS USUALLY HAS ABOUT 20 MEMBERS.

HOME AND HABITAT

If you want to see proboscis monkeys in the wild, you'll likely need to hop on a plane! Their home is the island of Borneo in Southeast Asia. That's the only place in the world they live.

These monkeys are arboreal, which means they live in trees. Their natural **habitat** is near the water. They occupy the **mangrove** forests along Borneo's coast or the **rain forests** along its rivers. Proboscis monkeys are also excellent swimmers and can travel easily on or below the water's surface.

STRANGE BUT TRUE! Proboscis monkeys have a special feature that makes them strong swimmers. Their fingers and toes are partly webbed, like the toes of frogs!

WHERE TO FIND THE PROBOSCIS MONKEY

THAILAND
CAMBODIA
VIETNAM
PHILIPPINES
SOUTH CHINA SEA
MALAYSIA
MALAYSIA
BORNEO
INDONESIA
INDIAN OCEAN
AUSTRALIA

MOST OF BORNEO IS PART OF INDONESIA, BUT SMALL PARTS BELONG TO THE COUNTRIES OF MALAYSIA AND BRUNEI.

TASTY LEAVES

You might've noticed proboscis monkeys not only have a large nose, they also have a big stomach! However, it's not because they eat a lot—it's because of what they eat. Proboscis monkeys eat lots of leaves. Leaves have **cellulose**, which is hard to **digest**. So, these monkeys have an unusual stomach full of special bacteria to help them digest the leaves.

Proboscis monkeys eat more than just leaves. They also eat fruit, flowers, seeds, tree bark, and bugs. Yum!

STRANGE BUT TRUE! A recent study suggests proboscis monkeys throw up their food and chew it again, like cows do! They're the only primates thought to do this.

PROBOSCIS MONKEYS EAT MOSTLY IN THE MORNING AND THE EVENING.

17

HUNTED

Sadly for proboscis monkeys, they have lots of enemies. They're hunted by many different animals. One of their main predators is the false gharial, a **reptile** that looks like a crocodile. Their list of enemies also includes clouded leopards, large lizards, giant snakes called pythons, and eagles.

People sometimes hunt proboscis monkeys for bezoar stones, or the mass of food and matter that hasn't digested in their guts. Why would hunters want bezoar stones? Some people believe these stones can help cure illnesses.

STRANGE BUT TRUE! People have other reasons for hunting proboscis monkeys. Sometimes they kill them for harming crops. Or, they might want to keep the monkeys as pets, even though this is illegal.

FALSE GHARIALS, SUCH AS THESE, ARE ABLE TO SEIZE ADULT MALE PROBOSCIS MONKEYS SITTING ON BRANCHES HANGING LOW OVER WATER.

ENDANGERED

Unfortunately, proboscis monkeys are endangered. This means they're in danger of becoming **extinct** because their population is quickly becoming smaller, and they're losing the habitat they need to survive.

People are destroying the habitat of proboscis monkeys. They have cut down trees where monkeys live for lumber and to make way for oil palm trees. Roads and houses are being built in the areas where the monkeys once lived. If people aren't careful, these special monkeys could disappear forever.

THESE UNUSUAL ANIMALS MAY LOOK STRANGE, BUT THEY MAKE THE WORLD A MORE INTERESTING PLACE. WHAT HAPPENS IF THEY DISAPPEAR?

GLOSSARY

bachelor: a young male animal without a mate

cellulose: matter that is the main part of the cell walls of plants

digest: to break down food inside the body so that the body can use it

extinct: no longer living

habitat: the natural place where an animal or plant lives

mangrove: a tropical tree that has roots that grow from its branches and that grows in swamps or shallow salt water

mate: one of two animals that come together to produce babies. Also, to come together to make babies.

primate: any animal from the group that includes humans, apes, and monkeys

rain forest: a tropical forest with tall trees that gets a lot of rain

reptile: an animal covered with scales or plates that breathes air, has a backbone, and lays eggs, such as a turtle, snake, lizard, or crocodile

webbed: connected by skin

FOR MORE INFORMATION

BOOKS

Gosman, Gillian. *Proboscis Monkeys*. New York, NY: PowerKids Press, 2012.

Owings, Lisa. *Proboscis Monkey*. Minneapolis, MN: Bellwether Media, 2014.

Zappa, Marcia. *Proboscis Monkeys*. Minneapolis, MN: Big Buddy Books, 2016.

WEBSITES

Amazing Monkeys: Why Do These Monkeys Have Such Outrageous Noses?
www.smithsonianchannel.com/videos/why-do-these-monkeys-have-such-outrageous-noses/56242
Watch a video about proboscis monkeys and their unusual noses here.

Nosey Monkeys
rangerrick.org/ranger_rick/nosey-monkeys
Find information and pictures featuring proboscis monkeys on this website.

Proboscis Monkey
www.nationalgeographic.com/animals/mammals/p/proboscis-monkey
Discover more about proboscis monkeys at this site.

Publisher's note to educators and parents: Our editors have carefully reviewed these websites to ensure that they are suitable for students. Many websites change frequently, however, and we cannot guarantee that a site's future contents will continue to meet our high standards of quality and educational value. Be advised that students should be closely supervised whenever they access the internet.

INDEX

babies 8, 10, 11, 12
bacteria 16
bands 12
bezoar stones 18
Borneo 14, 15
branches 8, 19
call 6
crops 18
day 12
face 4, 10, 11
false gharial 18, 19
females 4, 6, 8, 9, 10, 12
food 10, 16, 18
fur 10
groups 12, 13
habitat 14, 20
hunters 18

leaves 16
males 4, 6, 7, 8, 10, 12, 19
mating season 8
mouth 4, 6, 8
night 8, 12
mothers 8, 10, 12
nose 4, 5, 6, 7, 9, 16
oil palm trees 20
pets (illegal) 18
population 20
predators 18
shoulders 10
stomach 16
toes 14
water 14, 19
weight 8